The Origins of

The early history of a Nor

by

Roger Hawkins

The Morpathia Press,
13, Olympia Hill, Morpeth, NE61 1JH
© Roger Hawkins 2021
ISBN: 9781 902385 17 4

The path in front of this semi-circular seating area in Carlisle Park, or rather, its precursor in Anglo-Saxon times, is the place that gave Morpeth its name. The east side of Ha' Hill is visible in the background.

Notice the inspection chamber in front of the steps, indicating the culvert through which the Postern Burn runs under the flower park.

Introduction

The origins of Morpeth, and of its name, have engaged the interest of both historians and the inhabitants of the town since at least the 16th c. The Elizabethan antiquary William Camden believed that Morpeth was the Roman station of Corstopitum, but we may say at once that it was not, and that Morpeth has no Roman remains. This is not to say that there wasn't anything here in Roman times. There was a small Romano-British camp at Gubeon Cottage, and there are two much larger such camps at Park House Banks – which, however, are only visible from the air – but nothing that we would recognise as a town or village.

The town itself was founded in the 12th c. In 1188 or thereabout, Roger de Merley II issued a charter granting rights and privileges to "my free burgesses of the town of Morpathia", and in 1199 King John granted him a weekly market on Wednesdays and a fair on 22 July, the feast of St. Mary Magdalene.

This was the Morpeth that we know today, a planned new town on a greenfield site. Our object in this booklet is to look rather at the origins of the earlier settlement that stood somewhere near the parish church of St. Mary.

The chapters in this booklet first appeared as a series of articles in the *Morpeth Herald*, mostly in 2012. I have carefully revised and rewritten them, and I hope you will find pleasure in reading about the earliest history of our town.

Roger Hawkins February 2021
Morpeth

Chapter 1
Morpeth, the origin of the name [1]

Place-name studies are dry things. Here is the entry for Morpeth in Allen Mawer's *The Place-Names of Northumberland and Durham*:

Morpeth. c. 1200 Joh. Hex. *Morthpath*; 1199 R.C. *Morpeth*; 1210-2 R.B.E. *Morpat'*, c. 1250 T.N. *Morpath*; 1346 F.A. *id.*, *Morepeth*, 1428 *Morepath*.

Mawer concludes that the original form of Morpeth was the Old English morð-pæð, murder-path, "from some forgotten crime." This rendering, despite sounding like a tabloid newspaper headline, is the commonly accepted one, though for reasons I give in a later chapter, I think it is incorrect.

The staccato style of a place-name dictionary belies the interest of the documents themselves. One such, and a rather curious source, is the poem, *Esturie des Engles* (History of the English) by Maistre Geffrei Gaimar, written about 1150. This is 50 years earlier than Mawer's earliest source, though the four extant copies are all later. Gaimar was Norman-French, probably a priest, probably in Lincolnshire, and wrote it for a lady called Dame Constance. Here, Robert Mowbray, earl of Northumberland, has rebelled against William II:

> Li reis od son ost i alad,
> Le Nouel chastel idonc fermad;
> Puis prist Morpathe, vn fort chastel
> Ki iert asis sur vn muncel;
> Desur Wenspiz assis estait,

---

1. This chapter first appeared in the *Morpeth Herald* of 12 July 2012.

> Willame de Morlei laueit.
> E quant il out cel chastel pris,
> Auant alat en cel pais;

This means: "The king with his host went thither. The new castle then he built (this was Malvoisin, not Newcastle.) Then he took Morpeth, a strong castle, which stood on a hill; above Wansbeck it stood, William de Merley held it. And when he had taken it, he advanced into the country." This is from the British Museum's manuscript, which the editor regarded as nearest to the original. The Lincoln manuscript has Morpade, and Durham Morpape, but he confidently saw in this a misreading of Morpaþe, i.e. Morpathe.

Ha' Hill – vn fort chastel ki iert asis sur vn muncel

Here is an extract from Mawer's first source, the Chronicle of John, Prior of Hexham, written c. 1200. "In the same year [1138] on January 5th, a certain powerful man in Northumbria admitted into his property near the fortress called Morpeth (apud castrum quod dicitur Morthpath) eight monks from Fountains, who built a convent, called, as is well-known, Newminster (Novum-monasterium.)"

Hexham was not in the bishopric of Durham, but was a peculiar of the archbishop of York. In November 1313, Archbishop Greenfield wrote to the prior of Hexham. He has been examining the prior's fellow-canon, Brother William de Morpeth (Fratrem Willelmum de Morthpath, concanonicum vestrum.) We do not know what William stood accused of. It must have been serious if it could not be dealt with at Hexham, but whatever it was, he denied it.

William de Morpeth was a canon at Hexham

The archbishop sent him back to Hexham with orders to keep him safe until he could prove his innocence, but William was a desperate man; he escaped and went on the run, and on December 16th the archbishop wrote to the bishop of Durham to say that he has excommunicated Brother William de Morpath. He asks the bishop to do the same in his diocese, but especially around Morpeth (et specialiter apud Morpath.) William was at large for over six months, but was caught and taken to Bridlington priory, where he was kept for six months before being sent back to Hexham. The archbishop took care, after this second offence, to make William swear not to harm him or any of his people.

There is a rental of the lands belonging to the priory, called the Black Book of Hexham. It contains a much later reference to Morpeth, of 1470. In it, English and Latin are so mixed that you can almost read parts of it straight off, as in this account of Temple Thornton: "Et super lez Smal-half-acre, quia in medio, ex parte austr. (south) del Morpeth-way, ij acrae et di (2½ acres). Et apud le Propp, ex parte austr. de Morpeth-way, iij rodae (3 roods)." The entry for Neuton in Cookdale (Coquetdale) juxta Harbottel also has, incidentally, "ex parte le lonyng ibidem, inter toft' et croft' Johannis de Berehalgh". John may have been a Morpeth man; Berehalgh was a field where Bridge Street is now.

In the Red Book of the Exchequer, Mawer's R.B.E., the references to Morpeth are in long lists of scutages. Scutage was a money payment to the king in lieu of a knight's service. On p. 563, for the years 1210 to 1212, we have: "Norhumberlande. ... Rogerus de Merlay, baroniam de Morpat[h], per iiij milites." In other words, Roger was obliged to provide the king with four fully equipped knights, to serve for forty days in the year, or a sum of money in lieu, as a condition of holding his barony of Morpeth. On p. 713, in a summary of the amounts due to the exchequer for 1264-65, we have: "De baronia de Morpathe, xxvs. vjd." That is, the knights' services have been commuted for 25s 6d.

Mawer's T.N. of c. 1250 is the *Testa de Nevill*. The name means "Nevill's head", from a caricature of one of the exchequer clerks on the box that held the documents, but its proper title is Liber Feodorum, meaning the Book of Fees. It gives much the same information as R.B.E., but Mawer had only the 1807 edition, which is bad for dates. If you look in Hodgson's *History of Morpeth*, p. 7, you will notice that he dates these two entries to 1219 and 1240. There was a new edition of the *Testa* in 1920, in which the entries for 'Morpath' are dated 1212 and 1242-43, so Hodgson was very near the mark.

The register of Bishop Kellawe of Durham dates from 1311 to 1316. It has a number of references to Morpeth, always spelt *Morpath'*. I suppose the apostrophe is instead of proper Latin endings. The earliest is for October 1311, regarding an inquisition into the church at Morpeth. Soon after that, on January 12th 1311-12 (1312 to us, but 1311 until the end of March to them) a man called William de Bereford was collated to the rectory of the parish. William was ambitious, and in February 1312-13, the bishop gave him permission to take time off to attend the Schools. 'Morpath' also occurs in the register as a surname. On March 2nd 1312-13 Peter de Morpath was excommunicated for contumacy. More satisfactorily, on May 19th 1313, Ricardo de Morpath, rector of Graystok, was a commissioner to try a petition of Master William, vicar of Felton. Also in 1313, we have: "domino Reginaldo de Morpath, cappelano, de sex marcis receptio ab eodem de sequestro ecclesiae de Morpath'." Master Reginald the chaplain, apparently acting as agent for William de Bereford, has paid six marks to the bishop. Then, on August 3rd 1314, we meet an old acquaintance; the Bishop has finally excommunicated "frater Willelmus de Morpath', canonicus monasterii de Hextildesham" for contumacy.

ಬಿ ಬಿ ಬಿ

Further reading: Hodgson's *History of Morpeth* is widely available. Gaimar, the Book of Fees or *Testa de Nevill* (Part 1, p. 201), and odd volumes of the Red Book of the Exchequer and the Register of Richard de Kellawe, are available on the Open Library website. The Black Book of Hexham is available at Woodhorn, ref. LR271.

ಬಿ ಬಿ ಬಿ

Chapter 2

Early Morpeth, the village [2]

The early history of Morpeth is short on facts and long on supposition. The only grounds, for instance, for believing that the first settlement was somewhere near St. Mary's church, is the church itself. No remains of a village have ever been found. The building is of the 14th c., but the fact that it stands three-quarters of a mile from the town suggests the presence of an earlier village; and in 1984 the late Dennis Briggs discovered, by dowsing, the remains of an apsidal church (one with a semi-circular east end) underneath St. Mary's. Its walls were where the arcades are now, and the apse projected about 15 feet beyond the present chancel arch. The main axis of the church, Mr. Briggs says, "is some 22 degrees north of east, which suggests possible Anglo-Saxon origins. The plan recovered by dowsing could be late Anglo-Saxon or early Norman."

We do not know the name of the Anglo-Saxon farming village beside St. Mary's, except that it is unlikely to have been called Morpeth. I prefer to think of it as proto-Morpeth. Four place-names do exist for this part of the parish: Rectory, St. Mary's Field, Kirkhill, and High Church, but information on them does not go back very far.

Rectory speaks for itself. The 'Vicarige' is marked on Haiwarde's map of 1604, directly opposite the church, and with a large glebe adjoining it on the south side. It seems likely that it was there in earlier times as well. In Parson and White's *Directory* of

---

[2]. This chapter first appeared in the *Morpeth Herald* of 19 July 2012.

1827, the address of the Sun Inn is given as "Sun, John Robertson, Rectory, Tranwell."

The Church of St. Mary the Virgin, Morpeth

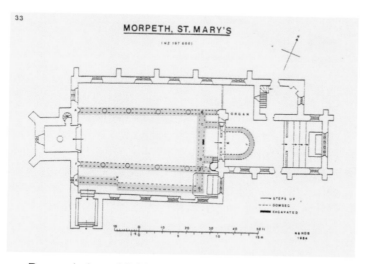

Dowsed plan of St. Mary's church by Dennis Briggs

St. Mary's Field is the hillside that lies to the north of the churchyard. It is an attractive site, facing south and bathed for hours with wintry sunshine on a clear December day. You could water your stock in the Churchburn at the bottom of the hill, and I should guess there were springs all over the hillside to provide water for cooking and brewing. Taking all of these factors together, St. Mary's Field is as likely a spot as any for the site of the original village.

St Mary's Field – as likely a spot as any

Kirkhill is the name of the large housing estate that extends nearly a mile west of the church. The name is said to have referred originally to the hill on which the church stands, but I doubt if this is correct. If you view St. Mary's from the bus shelter opposite the Sun Inn, you can see that it might well have stood on a fairly prominent bluff in primitive times, but it is by no means a significant feature, merely a low terrace overlooking then houses of St. Mary's Field. There is also a noticeably sharp rise from the road to the church, but it is purely artificial, having been created by wheeled traffic and by road improvements in the 19th c. and as recently as the

1920s. The conclusive evidence, however, comes from Haiwarde's map of 1604. It shows Kirkhill as a large field west of the church and quite separate from it, where Storey Park and the adjoining parts of the housing estate are now. [3]

St. Mary's stands on a low hill

In modern OS maps, High Church is simply an alternative name for the Kirkhill estate; but in the late 18th c. it was the name of a hamlet beside the turnpike from Newcastle, which included both the rectory and the Sun Inn. There were also some cottages on the roadside at the edge of Morpeth Common, but all that remains of them now is the clubhouse of Morpeth Golf Club, once the common herd's house. Whalton Road used to come out where the Club's drive is, and this former road junction is another likely spot for the one-time Anglo-Saxon village. If so, it suggests the possibility that

---

3. 'Kirk' is a common place-name element, e.g. Kirkheaton and Kirkharle. The late Harry Rowland speculated that the old village might even have been called Kirkhill.

Morpeth Common may originally have been part of its open fields.

Morpeth Golf Club is another possible site for the village

Hodgson says that Morpeth Corporation had no title deeds for the Common, having owned it time out of mind. And although it is no more than a hunch, one can imagine a situation in which, over a period, the more ambitious peasants would have left the village for the new town, leaving the ones that stayed to cultivate less and less of the open fields until what is now the Common reverted to rough pasture.

As for the name of our presumed Anglo-Saxon village, since its nearest neighbours (Hebron excepted) were Stannington, Whalton, Tritlington Choppington and Bedlington, it is quite likely that it too might have ended in 'ton' or 'ington', but we shall never know. I am tolerably certain that it would not have been called Morpeth until after the arrival of the Normans, possibly as early as 1080, or perhaps not until the reign of Henry I (1100-35).

For reasons we will go into later, I think the new name would have been applied first to Morpeth Castle, then to the Barony, and only afterwards to the village and, towards the end of the 12th c., to the newly formed town. [4]

༶ ༶ ༶

Further reading: John Hodgson, *History of Morpeth*, 1832. H. Dennis Briggs, *Hidden Churches of Northumbria*, 1987. For a discussion on the date of the founding of the Barony of Morpeth, see William E. Kapelle, *The Norman Conquest of the North*, 1979, pp. 193-4.

༶ ༶ ༶

---

4. William Kapelle says, "With the probable exceptions of Bywell and Callerton and possibly of Morpeth, none of the baronies in Northumberland originated before 1100."

Chapter 3
Crossing the Wansbeck [5]

The late Alec Tweddle once gave a lecture to Morpeth Antiquarian Society on river crossings in Morpeth, and there are surprisingly many: Low Ford Bridge, Skinnery Bridge, Bakehouse Steps, Oldgate Bridge, Elliott Bridge, Chantry Bridge, Telford Bridge, Stobsford Bridge, and of course the Stobsford itself. Downstream from that is the Quarry bridge, and finally the railway viaduct.

Old maps show other crossings, now gone. In 1860 there were stepping stones between Goosehill and the Terrace car park. Oldgate bridge was a suspension bridge with a ford next to it for vehicles. There used also to be a ford on the upstream side of the Chantry bridge, which for centuries was the recognised crossing place until the medieval bridge was built, and there was another ford below the Stobsford that served Borehole Lane.

Most of them were purely for local traffic. The crossing place that gave Morpeth its importance was the one near the Chantry, originally a ford, then the narrow, hump-backed medieval bridge, and now the Telford Bridge.

ಬಿ ಬಿ ಬಿ

If we consider the ancient geography of the town, the first thing to note is that the Wansbeck is not navigable and is an obstacle to travel. Secondly, the road from Newcastle to Alnwick and beyond is not Roman. What eventually became the Great North Road was no more than a succession of tracks linking one hamlet with the next in Anglo-

---
5. First published in the *Morpeth Herald* of 30 August 2012.

Saxon times. If so, it is interesting to speculate on why long-distance travellers in those days crossed the Wansbeck here rather than somewhere above or below.

Not crossing downstream is easily explained. The Wansbeck enters its gorge below Morpeth, and after that becomes tidal. From the earliest times, therefore, the ford at Morpeth was the lowest reasonably safe crossing place.

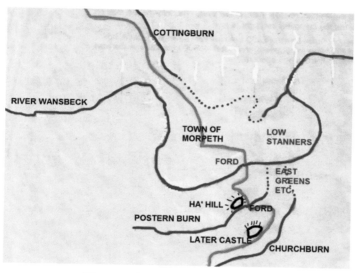

The crossing of the Wansbeck.
(Dotted lines indicate burns now culverted.)

The medieval bridge was built in the 13th c., but the ford may have had a bridge or submerged causeway even before that. The early form of Wansbeck was Wenespik. In *Northumberland Place-Names*, Stan Beckensall suggests that this could be Wain-Spik, i.e. Wagon-Bridge, 'spik' being an obscure word in both Dutch and German meaning a bridge made of tree trunks or brushwood. There appears to be no equivalent word in the

Anglo-Saxon, but it is suggestive that 'spæc' means a small branch or thin twig, and 'spelc' a splint, which has an obvious affinity with the Northumbrian spelk.

Until the mid-19th c. Morpeth was the lowest bridging point on the Wansbeck. Geographers take it for granted that everybody wants to cross a river at its lowest bridging point, but there is no obvious reason not to cross further upstream. The Roman road called the Devil's Causeway crosses the Wansbeck at Marlish, the Hart at Hartburn and the Font near Netherwitton, and another Roman road, Dere Street, now the A68, was one of the most travelled routes into Scotland right up to the days of turnpike roads. But one reason for long-distance travellers in Anglo-Saxon and Norman times choosing to cross at Morpeth rather than further inland would have been because it is on the direct route north from the Roman bridge at Newcastle, the Pons Aelius. Another, allowing Dere Street to be an exception to the rule, is that they would have encountered several large tributaries if they crossed further upstream.

ಬಿ ಬಿ ಬಿ

Having chosen to cross the Wansbeck somewhere near Morpeth, the question arises: Where exactly? It will be easier to see what options our travellers had if we imagine them coming from the north. After fording the Coquet at Felton, they crossed a belt of high ground and descended gradually to Bullers Green.

At that point their route became a narrow track with a steep drop on either side, the Cottingburn on the left and the Wansbeck on the right. This track, now Newgate Street, is a classic example of geographical determinism – no other approach to the river crossing would have made sense.

Newgate Street was a narrow track

Standing on the low terrace where Bridge Street is now, our travellers faced difficulties. The place where they stood was alright, but the south side was all obstacles.

On their left, it is likely that both banks were occupied by dangerous marshy areas, represented today by the Low Stanners on the north bank and the West, Middle and East Greens on the south. Both sides are protected now by flood walls.

There is no doubt about the marshy character of the south side; an abandoned river meander runs behind the houses, and serious flooding has occurred there within living memory. The north bank has also been flooded, as when Morpeth Library was damaged and its local history reference collection destroyed in 2008, but its ancient state is less certain. The Cottingburn enters the Wansbeck through here, but the area is on the whole well above river level; it would require a proper geological investigation to establish whether it is natural or built ground.

Dangerous marshy areas

On our travellers' right, the bank curved round in a forbiddingly steep cliff where the promenade is now.

A forbiddingly steep cliff where the promenade is now

The only safe option was a ford, just above the

present Chantry bridge. There an island-like hill, occupied now by Wansbeck Street, gave a safe foothold on the south bank.

We can no longer reconstruct the natural landscape south of Telford Bridge in detail, but a number of burns, now culverted, made their way into the Wansbeck somewhere near it. The main ones were the Churchburn and the Postern Burn. The land now occupied by the flower park and the courthouse, and extending eastwards as far as Crawford Terrace and Stobsford bridge, was a deltaic area where sluggish streams wound in and out between swamps and low-lying islands.

The only safe crossing place

So, not wishing to drown in a quagmire, travellers from the north crossed by the ford, gained the south bank near Wansbeck Street and turned right into Hillgate. They then made their way over the shoulder of Ha' Hill, crossed the Postern Burn as best they could, then climbed the path on the opposite side – now an obscure garden path – to where the War Memorial is now.

Over the shoulder of Ha' Hill

An obscure garden path – and a stiff climb

From there they followed the tarry peth to St. Mary's church. This ancient routeway is easy enough to follow, but does involve a stiff climb.

One might reasonably ask why they didn't follow the course of the modern road from Castle Square? The answer is simple: it didn't exist.

The Flower Park and Courthouse – a deltaic area with sluggish streams

The tarry peth

If you look carefully at the road between the Courthouse and Mafeking Park, it is actually in a cutting. The Churchburn at that point lies in a deep valley, which you can easily see if you peer over the stone wall at the side of Mafeking Park. Our ancient travellers did not want to go along the bottom of the burn, so they took the path over the hill instead.

The Churchburn gorge used to continue upstream, i.e. southwards, from Mafeking Park. Engineers in the early 19th c., probably Thomas Telford and his assistants, cut a ledge into the valley side to carry their new turnpike road. It was carefully graded so that the horses pulling the mails could go at a fast trot in both directions. At the back of the ledge, still visible from the Sun Inn to the Park Gates, they built a retaining wall of creamy sandstone.

In the 20th c., the burn above Mafeking Park was culverted. Its gorge was filled with the contents of the town dustbins, and Deuchar Park laid out when the landfill reached the level of Telford's road. It is difficult to picture this formidable valley as it was a thousand years ago, but once you can see it in your mind's eye, you can appreciate why the ancient trackway took the route it did.

ଛ ଛ ଛ

The Churchburn runs under Deuchar Park

Chapter 4
Where is Morpeth? [6]

In previous chapters we discussed possible locations for the original village and looked at the earliest documentary references to Morpeth. Different spellings occur in these old documents, but in all cases we have a two-syllable word of which the first part is 'Mor' or 'Morth', and the second part 'path' or 'peth'. Two documents, the earliest written c. 1200, have 'Morthpath', but most spell it 'Morpeth' or 'Morpath', in all cases with minor variations.

The first person to suggest an interpretation was William Camden, the Elizabethan antiquary. He suggested that Morpeth might be the Roman station of Corstopitum. Corstopitum, he thought, got altered to Morstopitum and later to Morpeth. He was a man of great learning, and the derivation was not implausible at the time. A century later, however, John Horsley, the Presbyterian minister in this town and author of *Britannia Romana*, quietly ignored it. It was left to John Hodgson in 1832 to point out Camden's mis-identification.

The most ingenious theory was that of G. Kennedy, in the *Story of Morpeth Grammar School*. He proposed the Celtic Mor-Beth – Great Burial Place, referring to a cairn on the ridge to the west of Ha' Hill excavated by William Woodman in 1830. "It consisted," wrote Woodman, "of a quantity of stones piled together; and appeared to have been one of the rudest description." Certain larger stones were

---

6. First published in the *Morpeth Herald* of 6 September 2012.

clearly the cap-stones, beneath which he found a thin layer of black earth, some fragments of bone, and nearby a piece of unglazed coarse pottery – hardly a great burial place!

The 'path' or 'peth' element is uncontroversial. The two words mean the same thing, but 'peth' has a special significance in Northumberland where it often refers to a stretch of road running steeply down to a river, as at the Lion Bridge in Alnwick, on the West Thirston side of the Coquet at Felton, and at Wooler. Felton has a Peth-foot, Durham City a Peth-bottom, and several places are called Peth-head.

All of these names arise from a landform that is everywhere around us: a river or burn in a deep valley. We are largely unaware of them because of the wonders of civil engineering. Whorral Bank, for instance, is steep, but would be a lot worse if it were not for the embankment over the Howburn. On a smaller scale, motorists coming into Morpeth from the south are oblivious of the Church Burn as they cross it just before Kendor Grove, yet in ancient times this little ravine must have presented a tricky obstacle to travellers of every kind. In short, until the days of turnpike roads, you couldn't go far without encountering a peth.

'Mor' or 'Morth', by contrast, are not the same at all. If Allen Mawer, the chief authority on place names in Northumberland and Durham is right, then Morpeth was originally 'Morth-peth', from the Anglo-Saxon *morþ* – the murder-path or death-path. My late friend Harry Rowland had his own variation on this theory, pointing out that Morpeth is first recorded only after the Norman Conquest; and since post-Conquest documents were written by French-speaking clerks, the root could equally be the

French *mort*.

John Hodgson proposed the obvious meaning: the town on the path over the moor. I think (no pun intended) that he was on the right track. With all deference to Mawer's expertise, the weight of his own evidence favours *Mor* as the first part of the name, not *Morth*, and the earliest reference to Morpeth, in Gaimar's *Esturie des Engles*, spells it Morpathe. It is true that we only have later copies of Gaimar's poem, not the original, but those copies, while differing from each other in other ways, consistently spell it Mor-.

Another reason for doubting the murder-path theory is that place-names commemorating a single event are rare. There must be others, but the only one I can think of is Battle in Sussex. Place-names are more often based on some permanent feature, whether natural or human. Permanent does not mean everlasting. You might look long for geese at Gosforth or nuns at Nunriding, and not find them, but they survived long enough to fix the name. And since moors are more or less permanent features of the landscape, I much prefer Hodgson's explanation.

But where was the moor and where the path?

To find the answer, we have to take account of both ancient geography and the ancient meanings of the word 'moor'. When we think of moorland today, we generally think of somewhere bleak, high and windswept, but in Anglo-Saxon times it simply meant land unfit for cultivation; it could equally be fen or marshland. A moor in the north of England is most often a bleak upland, while in the south it is more likely to denote a low-lying, marshy area. But neither rule is invariable, and both Morwick and

Morralee are thought to incorporate the idea of a lowland moor.

There were certainly moors at Morpeth. In 1239, Roger de Merley III gave permission for his burgesses of Morpeth to dig turf in his turbaries at a penny a cartload – a clear indication of the existence of a lowland moor. And in 1389 the men of Mitford and Morpeth both dug turves on a piece of land in dispute between them, known as Threpmore.

We owe Morpeth as a place-name to the military genius of the Normans. My guess is that it was first applied to the castle, and thence to the barony. Whatever the village at St. Mary's was called, it now took the name of Morpeth, as did the later town founded in the 12th c.

To understand how this happened, let us imagine a traveller, this time coming from the south. He follows the old road uphill from St. Mary's, and descends by a dangerous slope to the Postern Burn where it emerges from its steep-sided little valley.

Where the Postern Burn used to emerge

Our traveller would not want to force his way up the

ravine, and whereas nowadays the burn disappears into a culvert under the flower park, at that time it widened oozily into a dangerous marsh – a moor in the original sense of the word, that extended eastwards to Stobsford Bridge. Our traveller had no option but to cross the burn where he stood. It was a classic pinch-point, and the wily Norman, probably William de Merley I, built his castle on Ha' Hill, directly above it.

Ha' Hill controlled a pinch-point at the head of a dangerous marsh

The baron would naturally want to know what the place was called where he had chosen to build his castle, and one can well imagine that this would be a difficult question for the local Anglo-Saxon peasants to answer.

If you think of someone nowadays asking where one is on a motorway, the most realistic answer is that you aren't anywhere. You're just on the motorway. It was the same with this place. The village was half a mile behind, and the ford two hundred yards ahead.

His honour's castle was simply on the path where it skirted round the moor – the Mor-peth.

ଛ ଛ ଛ

Acknowledgement: The idea behind this chapter arose from a lecture given several years ago by Christopher Hudson, when he remarked that property boundaries point to the ancient King's Highway going along Hill Gate, not through Castle Square.

ଛ ଛ ଛ

In the valley of the Postern Burn

ଛ ଛ ଛ

The valley of the Postern Burn today, from the side of Ha' Hill – A traveller would not want to go there

The sides of the valley are both steep and high – View over the Burn from Ha' Hill towards Morpeth Castle

Chapter 5
William of Morpeth sells his fields [7]

In his *History of Morpeth*, 1832, John Hodgson's records a number of land transactions between ordinary residents of the town, dating from the late 13th c. The people engaged in these transactions, some of whom were women, were not your traditional medieval peasant. They were not bound to the land and obliged to work certain days every week on the lord's demesne, but free tenants who paid money rents for the lands they occupied. Their personal obligations were limited to "the accustomed services to the lord", which seems in practice to have meant attending the twice-yearly manorial courts in person.

Some of these transactions related to burgage plots in the town itself, but others were about land in the surrounding area. As an example, under a deed of 1283 or thereabouts, William of Morpeth conveyed to Richard de Morpeth, Clerk, three acres of ground in the Field of Morpeth.

We don't know the exact boundaries of the Field, but other deeds show that it went as far south as Clifton. It seems to have extended across to Stobhill (there was no East Coast railway line in those days) and probably included part of what is now Morpeth Common.

William's property was in a part of the Field called the Florys. It consisted first of an acre called the Shorteaker, lying between the lands of Robert of the Park on the east and Robert at Church on the west. The other two acres were made up of eight roods, a

---

7. This chapter was first published in the *Morpeth Herald* of 30 April 2020.

rood being a quarter of an acre. There were five roods containing six selions abutting the land of Roger Cramper; one rood between the lands of Nicholas and Robert, near the Stadandstone; another extending to the Kyrckeburn, also between the lands of Nicholas and Robert; and one going towards Halleslath, near the land of William son of William. [8]

The pieces of land in question are defined very precisely, leaving no room for doubt about their location and extent – except that we don't know where the lands of Robert of the Park, Robert at Church, Roger Cramper, Nicholas and Robert, or William son of William, were. Nor, with the possible exception of the Kyrckeburn, do we know where the physical landmarks were.

As regards the local landmarks, the Florys looks as if it ought to refer to flowers; but the Anglo-Saxon word *flor* means floor, foundation, courtyard or threshing floor. If this is the meaning, the Florys must have been part of the Field where a settlement had formerly stood, leaving behind the remains of floors, foundations and cobbled yards. If so, it would have been close St. Mary's church. [9]

Kyrckeburn is obviously the Churchburn. It starts on the golf course, just to the south of Whalton Road

---

8. A selion was sometimes a measure of land equal to an acre, but in this instance it means only that the block of five roods (1¼ acres) was divided into six strips.

9. The Anglo-Saxon or Old English for flower is *bloma*, as in our word *bloom*. *Flower* is a Middle English word that came in from the French after the Norman Conquest. By the late 13th c., everyday language would undoubtedly have been Middle English, but the names of landscape features that had existed from time immemorial would still be essentially Anglo-Saxon.

ends, passes under the South Road and round the south side of Kendor Grove, then past the back of the Cottage Hospital site, after which it winds through Rectory Park to emerge at the side of the road near the Sun Inn. It then disappears underground in its culvert beneath Deuchar Park.

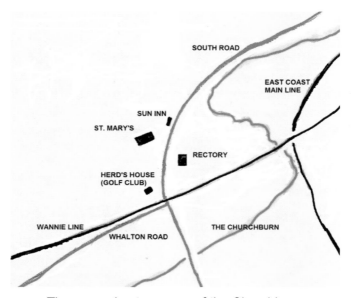

The approximate course of the Churchburn based on the OS 6-inch map of 1866

The part of the burn's course behind the Cottage Hospital and through Rectory Park was all glebe land, i.e. it belonged to the rector, so the rood extending to the Churchburn could not have been there. [10] For reasons touched on in Chapter 3, I

---

10. Haiwarde's map of 1604 shows the glebe very clearly. Haiwarde marked what I take to be Whalton Road in a half-hearted sort of way, but the Churchburn not at all. It is difficult to be sure, therefore, but it looks as if the southern boundary of the glebe corresponded either with that of Kendor Grove, or with the burn itself a few yards to the south.

don't think much of the burn's course under Deuchar Park would have been attractive as farm land. This leaves three areas where it might have been: viz. near Whalton Road ends (i.e. a little way south of Morpeth Golf Club's clubhouse), or just south of Kendor Grove, or close to the Sun Inn.

Halleslath is made up of two words, *halles* and *lath*. Of these, *lath* is the easier, being derived from the Anglo-Saxon *læð* (pronounced *lath*, with a flat 'a') meaning land. There is a slight suggestion in Bosworth-Toller's *Anglo-Saxon Dictionary* that it might occasionally mean a lee or meadow, but the main meaning is 'land'. *Halles* has three possible derivations: from *hal*, meaning an angle, corner or secret place, or *heall*, actually two different words, one meaning a hall or residence and the other a rock or stone.

Even experts in place-name studies sometimes have to choose which of several possible meanings to adopt, and I take Halleslath to mean the land belonging to the hall. If so, it would appear to have been the land belonging to an important house, or perhaps to the site of a former such house, which had been the residence of the village head man in the days before the Conquest.

The Stadandstone is another fascinating word. It looks as if it derives from the Anglo-Saxon *staðol*, or *stathol*, a foundation, hence the southern English staddle-stone. If so, it would have been the survivor of an arrangement of stones that once supported a granary or similar building, and would make an obvious landmark.

All of these pieces of land, remember, were in the Florys. We have to be careful here of two things: one is circular reasoning, and the other the tendency to take assumptions as ascertained facts.

The only real facts we have are, first, that St. Mary's church existed in Anglo-Saxon and early Norman times, and secondly that William of Morpeth's three acres that he sold to Richard were all in the Florys.

As for St. Mary's, we can be sure that a church stood on that site in the immediate post-Conquest period, say about 1080, and almost certainly before that in Anglo-Saxon times. As for the pieces of land being in the Florys, we can only assume that the deed would have been drawn up with all proper care.

There seems no reasonable doubt that the Florys with, as its name implies, foundations, floors and yards, was the remains of a settlement. This impression is strengthened by the evidence of the two landmarks, the Stadandstone and Halleslath. As with the Florys, everything depends on the meaning of place-names. An expert in such matters might interpret them differently.

The reference to the Churchburn is vital because it tells us that the Florys was close to the burn. If so, there are only two or three places where the Florys could have been, either just north or just south of the church

Subject to all of that, we can now draw three conclusions: First, that the Florys was the site of the old village; secondly, that the village had been abandoned by the 1280s, though some dwellings, notably the rectory, still existed nearby; and thirdly, that our earlier assumption, that it was either on St. Mary's Field or near the golf club, was correct.

The town of Morpeth, founded c. 1188 – see p. 3

Newgate Street from the River Wansbeck – see p. 18

The site of the Anglo-Saxon ford – see p. 20

Retaining wall for the turnpike road – see p. 24

Available from:
13, Olympia Hill, Morpeth, NE61 1JH
£6.00, post free in the UK
Cheques payable to Roger Hawkins please
Email: rhcu@btinternet.com

Printed in Great Britain
by Amazon